Passive Income

Highly Profitable Passive Income Ideas On How To Make Money Online And Start Your Own Online Business

Anthony Parker

ISBN:1981334963
ISBN-13:9781981334964

Introduction

Thank you for taking the time to purchase this book: *Passive Income: Highly Profitable Passive Income Ideas on How to Make Money Online and Start Your Own Online Business.*

This book covers the topic of generating a passive income and will teach you everything you need to know about starting your own online business.

At the completion of this book you will have a good understanding of multiple ideas you can use to create a passive income and be able to make money online through your own online business.

Once again, thanks for purchasing this book, I hope you find it to be helpful!.

TABLE OF CONTENTS

Chapter One: Introduction To Passive Income

The internet has brought about numerous opportunities for earning a decent income and living a flexible, independent lifestyle. The most alluring of these opportunities is the ability to earn a passive income. Everyone dreams of the ability to make money without trading their time for it. The ability to earn a passive income is high sought after, and unfortunately, it is also highly misunderstood.

What Is A Passive Income?

To most people, the words 'passive income' conjure images of money constantly flowing into their bank accounts while they sleep on the couch or relax on a beach in Bali. This is a somewhat warped view of passive incomes. Essentially, a passive income is an income stream that doesn't require a lot active maintenance or involvement to keep the money rolling in. While you are not actively involved in running the business, it doesn't mean you can ignore it either.

However, one thing most people don't understand about earning a passive income is that it requires a lot of upfront investment, either in time, money or both. During the initial stages, there are usually no returns from all the effort and investment put into a passive income venture. However, once the systems are up and running, a passive income venture can maintain itself, earning you money without a lot of effort.

In this book, am going to give you 7 highly profitable ideas that you can use to build your online business and earn a passive income. Let's dive in.

Chapter Two: Cryptocurrency Trading And Investing

Cryptocurrencies are digital means of exchange that are based on rules of encryption and cryptography. In recent years, cryptocurrencies have gained a lot of popularity as a new method of making money. According to coinmarketcap.com, cryptocurrencies have reached a total market capitalization of about $152 billion. Trading in cryptocurrencies is a venture that can be highly profitable for both beginners and novices. But what exactly is cryptocurrency trading?

Cryptocurrency trading involves trading different kinds of cryptocurrencies against each other. Think of it as forex trading, but instead of trading fiat currencies, you will be betting on the price movement between different cryptocurrencies, or between cryptocurrencies and fiat currencies. The most common kind of cryptocurrency trading involves trading Bitcoin against US dollars or Bitcoin against other altcoins, something that can also be referred to as altcoin flipping. It's important to note that most altcoins cannot be bought via fiat currency. Cryptocurrency trading is a great way of earning a passive income because it has very little entry barriers and in some cases, it doesn't even require any form of verification.

Why Trade In Cryptocurrency?

Before we go into how you can trade in cryptocurrencies, let's look at why cryptocurrency trading and investing is such an attractive option.

Cryptocurrencies are global: Unlike fiat currencies, cryptocurrencies are not tied to the policies or economy of any country. This makes them an attractive investment option since they cannot be controlled by any government. However, this means that the price of a cryptocurrency will be affected by a wide range of world events.

Trading is open 24/7: Unlike the stock market, there are no official cryptocurrency exchanges. However, there are dozens of informal cryptocurrency exchanges all over the world that are in operation 24/7. This is beneficial because it allows you to make trades whenever the trends become favorable, without having to wait for the official exchange to open. The lack of official exchanges also means that there are no official prices for cryptocurrencies. This creates arbitrage opportunities, though the prices are generally within a similar price range for most exchanges.

Cryptocurrencies are highly volatile: Unlike stocks and fiat currency, cryptocurrencies experience rapid and frequent price fluctuations. This volatility creates great opportunities for traders to enter the market and make quick profits.

How To Trade Cryptocurrencies

Finding An Exchange

The first thing you need to do before you can start trading cryptocurrencies is to join an exchange. However, we have already noted that there are no official cryptocurrency exchanges. While there are multiple unofficial exchanges, choosing the right one can be a bit challenging, especially if you are a beginner. Before making a choice on the best exchange for you, you should consider the following factors:

Trust: How trustworthy is the exchange? Is there a chance that the exchange might run away with your money? To find the trustworthiness of an exchange, you should read online reviews from other people who have used the exchange.

Location: It is always advisable to go for an exchange that accepts deposits in your country's fiat currency.

Fees: What fees does the exchange charge? Cryptocurrency exchanges with lots of fees are going to eat into your profits.

Liquidity: If you intend to trade large amounts of cryptocurrency, you should go for a large exchange that has good market depth and high liquidity.

Some popular cryptocurrency exchanges you should consider include:

Coinbase: This is one of the most popular cryptocurrency exchange in the world. The platform supports Bitcoin, Ether and Litecoin. The exchange is accessible to users through both desktop and mobile devices. Coinbase has a great reputation and good security, is user friendly even for beginners and has reasonable fees.

Kraken: This exchange is the largest in euro volume and has the highest liquidity. Kraken allows users to trade between Bitcoin and fiat currencies like US dollars, euros, British pounds, Canadian dollars and Japanese yen. It also supports other cryptocurrencies like Ethereum, Ethereum Classic, Monero, ICONOMI, Dogecoin, Litecoin, Zcash, Ripple and many more. While Kraken has a great reputation and decent exchange rates, it is more suited to experienced traders.

Poloniex: This is the world's leading cryptocurrency exchange. Poloniex does not support fiat currency. However, it allows users to trade Bitcoin for over 100 types of altcoins. It has advanced tools that make it easy for traders to perform trade analyses. Poloniex is a great choice since it has low trading fees, high volume trading and is very user friendly.

Shapeshift: Just like Poloniex, Shapeshift does not support fiat currency. However, it supports multiple cryptocurrencies, including Bitcoin, Dogecoin, Dash, Ethereum, Zcash and Monero. One of the biggest advantages of Shapeshift is that it allows people to trade without having to sign up for an account. Shapeshift has a good reputation, reasonable prices and allows fast trading.

Bitsquare: This is an open source platform that allows users to trade Bitcoin for different cryptocurrencies as well as fiat currency. Just like shapeshift, one does not need to sign up for an account before accessing Bitsquare.

Figuring Out When To Buy And Sell

The most important aspect of trading cryptocurrencies is knowing when to buy or sell. Unfortunately, there is no direct answer to this. Just like forex trading, predicting the direction that a cryptocurrency is going to move is not an easy thing. You can never be certain. It is all based on speculation. However, there are some basic principles that you can follow to make more accurate predictions. You can learn to identify fundamental factors that affect the prices of various cryptocurrencies. For instance, if a popular cryptocurrency exchange starts supporting a new cryptocurrency, you can expect the price of the said cryptocurrency to increase. Governments can also affect cryptocurrency prices. For instance, in September 2017, China banned trading in Bitcoin, resulting in a temporary drop in the price of Bitcoin.

Some Important Cryptocurrency Trading Tips

To increase your chances of making profits with cryptocurrency trading, here are some few tips to keep in mind when trading.

1. Buy low, sell high. This is the golden rule, whether you are trading in forex, stocks or even real estate. Watch for cryptocurrencies that exhibit repeated patterns. Once you identify these patterns, buy the cryptocurrency when the price is low and sell when the price peaks. There are several cryptocurrencies that exhibit such patterns.

2. Make a point of thoroughly studying the charts. This makes it easier for you to detect bullish waves. Once you detect a bullish wave, you can then find a good entry point to buy. As the price rises, you can then find a good selling point. The more you hone this skill, the easier it will be for you to make profits.

3.   Coins that have high trading volumes usually tend to be very profitable, especially on an exchange like Poloniex. The opening of a new cryptocurrency on an exchange is usually followed by great increases in price for the first 2-3 days. This is a great time to make good money.

4.   Always maintain your common sense when trading. Do not put all your eggs in one basket. Instead of investing all your capital in one cryptocurrency, you should split it up into several parts and use each part to invest in a different cryptocurrency.

5.   Watch out for price variations of the same cryptocurrency on different exchanges. Sometimes, it is possible to buy a cryptocurrency at a low price on one exchange and sell it for a profit on another.

Buying And Holding Cryptocurrencies

If you find trading in cryptocurrencies a complicated venture, you might choose instead to buy and hold the cryptocurrency. For instance, the price of Bitcoin rose from about $1000 in

January 2017 to over $5000 in October 2017. If you had invested in the cryptocurrency at the beginning of the year, your money would have increased over five times.

There is no way of predicting the growth of a cryptocurrency, but if you feel bullish on the cryptocurrency, you can opt to buy it and hold onto it for a couple months or years. If the price goes according to your predictions, you can easily make some significant amounts of money without having to do much. However, you need to realize that this is essentially a gamble. Most people who make money through cryptocurrencies do it by banking on the volatility that occurs every day.

Buying and holding cryptocurrencies is more suited to beginners. Unlike trading where you must keep track of the charts throughout the day, there's not much you need to do if your intention is to hold. Simply buy cryptocoins, keep them safe and once the price appreciates enough, sell them and earn your profit. The most important thing to remember here is that there is no guarantee that a cryptocurrency will grow in value. Therefore, you should only invest funds that you can afford losing.

Risks Associated With Cryptocurrency Trading And Investing

While trading and investing in cryptocurrencies might be a great way of earning a decent income without a lot of effort, it is not without its risks. Some risks you should watch out for when trading cryptocurrencies include:

Leaving your funds on the exchange: One thing you need to understand with cryptocurrency transactions is that they cannot be reversed. This is why you should take a lot of care to ensure that you keep your coins secure. If someone steals them, you have no way of getting them back. To avoid the risk of your coins getting stolen, you should never leave them on the exchange. You should always be in control of your funds.

The collapse of Mt. Gox is a good illustration of why you should never keep your coins on the exchange platform. Mt. Gox was one of the largest exchanges during the early days of Bitcoin. Many Mt. Gox users, forgetting the importance of always being in control of your own cryptocurrency coins, left over 800,000 Bitcoins in their Mt. Gox accounts. In 2014, some hackers exploited a bug and stole these Bitcoins, leading to the eventual bankruptcy of Mt. Gox. All the users who had left their Bitcoins in their Mt. Gox accounts lost their funds.

To avoid losing your funds if an exchange gets hacked, you should *never* leave your cryptocurrency coins in the exchange.

Your capital is at risk: As with any other kind of trading, you might end up losing all your money if things go wrong. As a beginner, you should only invest small amounts as you learn the ropes. Alternatively, you can use a demo trading account until you are confident in your ability to consistently make profits.

If you are looking for an opportunity to generate a passive income without having to put in a lot of time and money in the beginning, you should consider getting into cryptocurrency trading and investing. Even if you are a complete beginner, this is something you can learn in a very short time and be on your way to making serious money.

Chapter Three: Blogging

Many people regard blogging as a mere hobby, a fad that can only earn you a couple of dollars at best. This view is not totally baseless. There are millions of live blogs on the internet, with many more going live each day. A huge percentage of these blogs do not bring in any significant incomes for their owners. However, did you know that there are bloggers who make millions of dollars through their blogs? The best part? You can too, if you have the right strategy and if you put in the effort.

The difference between running a blog that brings in $10 each month and one that rakes in $1,000,000 lies in your approach to blogging. Bloggers who earn serious incomes from their blogs know that their blog is not merely a tool for publishing and sharing their thoughts. They know that their blog is a business and they therefore treat is as one. They know that, just like building a business, building a successful blog takes a lot of time, dedication and learning.

In this chapter, I am going to take you through the basic steps you should follow to earn a passive income through blogging. Before we start, I want you to note one thing: blogging needs a lot of upfront work before you can start making significant amounts of money. It might take you about a year before you start making an income. Don't give up during this period.

With that out of the way, let's dive into the actual steps of creating a profitable blog.

Step One: Set Up Your Blog

This step is obvious. Before you can start making money through a blog, you need to have the actual blog. Setting up a blog is a pretty easy and straightforward thing. There are many platforms you can use to create your blog. My favorite platform is WordPress, since it is easy to use even for complete beginners and allows lots of customization. Before you set up your blog, you first need to come up with a blog name. Pick a blog name that is short, memorable and one that gives your readers an idea of what your blog is about. After choosing your blog name, now is time to register your domain and buy hosting. The domain is your blog's address, while the hosting is the place where your blog lives on the internet. If possible, use your blog name as your domain.

Once you have registered your domain and paid for hosting, simply set up your blog with a platform like WordPress, select a theme for your blog, set up the necessary pages and that's it. Now your blog is live and you are ready to start posting.

Step 2: Start Creating Useful Content

Creating a blog without content is like setting up a restaurant with tables but without the food and waiters. No one will come. After you set up your blog, you should focus on creating useful content for the blog. Content is what attracts readers to your blog. The kind of content you create for your blog will depend on the topic that your blog is focused on. To be successful as a blogger, you should create content that is geared towards a specific niche or demographic. If you try to write everything for everyone, you won't be able to attract a loyal audience, which will in turn curtail your chances of success.

When it comes to content creation, there is one key thing you should always have in mind. Your content should be valuable and useful. Always aim for content that will make people's lives better in some way. Doing this creates a strong relationship with your readers. It will make them like and trust you – which are two essential things you need from your audience if you intend to make money from your blog.

Another important thing you should adhere to is consistency. Create a content schedule and stick to it. If you decide that you are going to share your posts once a week or once a month, stick to that schedule. Being consistence helps you to create a following for your blog.

Step Three: Promote! Promote! Promote!

Creating content is one part of the equation. The other part involves finding people to read your content. Most newbie bloggers approach blogging with a "build and they will come" mentality. They think that once they create the content, readers will automatically find it. However, things rarely work that way. Remember, during the early stages of your blog, no one knows about it, except for a few friends. Therefore, even if you keep creating content for 5 years, without promoting the content, you will see very little growth.

The key to getting out of this rut is spending substantial amounts of time and effort building traffic for your blog. I recommend that you spend 20% of your effort on content creation and the remaining 80% on content promotion. Think of it this way. A blog with five posts and 10,000 readers has a greater impact than a blog with 50 posts and only 100 readers.

Before you start promoting your blog, think about the kind of people you want to attract to your blog. Create a reader persona or profile to help you figure out the kind of people you want on your blog. Then find out the places where you are likely to find such people and focus your efforts in promoting your blog on those places. To help you discover where to find your target audience, ask yourself questions like these:

- What are the top blogs they read?
- What kind of online forums do they participate in?
- On which social media networks are they more active?

Once you know the blogs your target audience likes to read, you can guest post on those blogs. If they participate in specific forums, join those forums and start engaging with them. Join the social media networks they are active on and promote yourself there. When building an online presence and promoting your blog, the key thing is to be helpful and to nurture relationships – not to spam people with links to your blog posts.

Step Four: Build Engagements

If you are consistent with your content creation and content promotion efforts, you will start seeing an increase in the number of people who visit your blog. Some will begin engaging with your content. This is the time you should start focusing on building a community around your blog. Respond to the comments readers leave on your blog and reach out personally to the readers if need be. The aim here is to create a community of loyal followers who will keep coming back to your blog again and again. Once your readers feel like they know you and that they can trust you, they will turn into evangelists, telling their networks about your blog, which will grow your blog even more. It is much easier to make money from a blog if you have a loyal and engaged community around your blog.

Step Five: Start Earning Through Your Blog

This is the point where most bloggers want to find themselves. However, you cannot get here without going through the first four steps. Those first steps are the foundation block for a blog that is capable of bringing in a consistent income for the long term. However, you should be aware that having implemented the first four steps properly does not mean that money will start rolling in automatically. You will still need to put in more effort and keep experimenting.

With that out of the way, let's now look at how to make money through your blog. There are several ways of monetizing your blog. These include:

Advertising

Advertising is the most popular method of earning through your blog. This is the method many bloggers use to make their first dollars from their blogs. Advertising through a blog can be compared to selling advertisements on magazines and newspapers. Advertisers are obsessed with numbers. This means that once your blog starts growing and attracting substantial amounts of traffic, advertisers will be more than willing to pay you to get their products in front of your readers.

When it comes to selling ads on your blog, it is more profitable to have direct deals with the advertisers. However, before you can land a direct deal, you need quite a huge traffic. Luckily, there are ad networks such as Google AdSense which allow you to publish ads on your blog even if you only have modest traffic. Apart from selling ad space on your blog, you can also make money from inline text ads and sponsored blog posts. You could also create video and podcast ads as well as job boards and share them on your blog.

If you are able to attract substantial amounts of traffic to your blog, selling ads is a great way of generating a passive income since it doesn't require a lot of hands-on effort. Simply put up the ad and wait to get paid.

Selling Products Through Your Blog

Today, we live in a world that is more concerned with consuming than producing. If you can produce something, there is a ready market of willing consumers. You can take advantage of this to make money off your blog. For instance, if you have some programming skills, you can take a common problem that your readers struggle with, build an app to help them overcome the problem and then sell the app through your blog. If you are knowledgeable in something that can help people solve a problem or become better at something, you can package that knowledge into an eBook and sell it through your blog. Alternatively, you can package that knowledge into an online course and sell it to your readers.

Digital products are not the only thing you can sell through your blog. You can use it to sell physical products as well. If you are a writer, you can write a paperback book and use your blog to sell it. Fashion bloggers can use their blogs to sell clothes and fashion accessories to their readers. The beauty of using your blog to sell your products is that you don't have to pay any commissions to anyone. You keep 100% of the proceeds from the sale of your products.

Selling Your Services

A blog provides you with the perfect avenue for showcasing your skill and expertise in a specific field. If your business is modelled around the provision of services, you can use your blog to build credibility for your services and attract potential clients. For instance, if your business provides digital marketing consultancy services, your blog should share helpful, informational tips about digital marketing. Your blog should act as a resource for anyone who wants to master digital marketing. In this case, if someone who has been reading your blog needs to hire a digital marketing consultant, you will be the first person on their mind.

This is a great model that can be adopted by anyone whose business is pegged on the provision of services. It is especially suitable for people who provide remote freelancing services, such as web design and development, copywriting, virtual assistance, etc.

Promoting An Offline Business

A blog is essentially a lead generation machine. It helps you find people who are interested in a certain topic or niche. You can then sell these people stuff related to their interest. It doesn't matter if whatever you decide to sell to them is from an offline business. An offline business can use a blog to reach an online audience. Think of it like an offline business marketing their products through a website or a social media network. Studies show that about 64% of shopper's research online before they commit to making an offline purchase. This means that with a blog, you have an awesome opportunity of promoting your offline business to your online audience.

Promoting an offline business through a blog also allows you to build your business brand, grow your business's credibility and authority and build a community of loyal customers around your business.

Events

While this method is yet to gain a massive foothold in the blogosphere, the number of people using their blogs to run events and make a ton of money is rising steadily. This method involves holding events such as live workshops, webinars and seminars, online summits and conferences. As a blogger holding such an event, you would charge people to attend the event, using your blog to market and promote the event. If you have been consistently sharing valuable nuggets through your blog, people will be willing to pay to attend your event if you promise to teach them something exclusive. Alternatively, you might opt to find sponsors for your event.

Apart from selling virtual tickets to your event, you can also make money indirectly by promoting your products and services to the people who attend the events. You could also invite guest speakers to your event and charge them commissions for the sales they make from selling their products and services to the attendees.

Recurring Revenue

This is one of the best ways of earning a passive income through your blog. This method involves finding customers and then having them pay for a product or service every month. This means that once you find a client, you will continuously earn from them every month without having to put in a lot of extra effort. It is also easier to maintain a recurring customer than to find new customers each month.

To earn recurring revenue from your blog, you first need to come up with a high value product or service that people are willing to pay for continuously. This could be an exclusive membership to a private community, one-on-one coaching or access to premium content. Once you have created your high value offering, you should then come up with a member's area or offer a subscription based model where only people who have paid have access to the high value offering.

Affiliate Marketing

This is another popular model of earning a passive income through your blog. More and more bloggers have started incorporating this method into their blog monetization strategies. Affiliate marketing simply means promoting another person's products and services to your audience. You then get a commission for every sale that results from someone you referred to the product. You can monetize your blog's traffic through affiliate marketing. I am going to discuss how you can do this in the next chapter.

Other Income Streams

The methods discussed above are some of the most popular if you want to earn a passive income from your blog. However, they are not the only methods available. There are numerous other ways of earning through your blog. For instance, you could ask your readers to make donations to your blog, based on the kind of content you share on your blog. You can also syndicate content to other sites and earn from that.

Another important thing to note is that if you want to earn a substantial income from your blog, you should not tie yourself to one monetization method. Instead, you should combine a variety of methods to diversify your income. This helps you spread your risk and allows you to start earning a full-time income from your blog much faster.

Chapter Four: Affiliate Marketing

Affiliate Marketing is the process through which a marketer gets rewarded for each sale or prospect brought to a business through the affiliate's marketing efforts. In most cases, affiliate marketers get a commission from every sale they drive. Affiliate marketing is a very easy, fast and passive way of making money online. This is because you don't have to spend any time, money or effort creating your own products. Instead, you only need to find customers for other people's products. Affiliate marketing is great for people who have access to a huge audience, such as bloggers, podcasters and YouTubers.

How Affiliate Marketing Works

When you join an affiliate program, you are issued a unique affiliate code which you then use to refer traffic to the seller's site or products. This allows the affiliate network to track all the sales that are made through your affiliate code. This code comes in a variety of formats, including text links, banners and other kinds of creative media. Once a prospective buyer clinks on one of your unique links, they are directed to the seller's site or product page. If they make a purchase, you earn your commission, simple and clean! You also get complete access to real-time stats of all your sales and accrued commissions.

With some affiliate networks, you don't even have to make a sale to earn a commission. The most popular payment structures for affiliate programs are:

Pay per Sale: This is the most common payment structure for most affiliate programs. You get paid a percentage of the sale price for every sale made through your affiliate code.

Pay per Click: In this payment mode, you get paid based on the amount of traffic you drive to the seller's website, regardless of whether they make a purchase or not.

Pay per Lead: With this model, you get paid once a person you referred to a seller's site provides their contact information.

Advantages Of Affiliate Marketing

Affiliate marketing is one of the best techniques of earning an online income. Some advantages of affiliate marketing include:

Cost effectiveness: With affiliate marketing, you can earn a substantial income without having to worry about producing a product, hiring a physical business location and having to pay employees. You don't have to pay for product storage, packaging and shipping either.

Wide reach: Since you will do most of your affiliate marketing online, you can reach prospective buyers from any part of the world.

Zero fees: Most affiliate networks and programs don't require any payment before you can join.

No customer support: As an affiliate marketer, you only refer prospective buyers to the product you are promoting.

Providing customer support and dealing with complaints and returns is up to the seller.

Passive income: Affiliate marketing is a good way of earning a passive income. You can generate a steady flow of income without having to be in front of your computer all the time.

Tips For Becoming A Successful Affiliate Marketer

Despite having such advantages, it doesn't mean that you will make overnight wealth through affiliate marketing. Affiliate marketing is highly competitive. There are multiple strategies for you to follow to become a great affiliate marketer:

Product Choice

Don't register with all the affiliate programs you come across and try to market a whole multitude of products. While it might seem like a great idea, it is counter-intuitive. Instead, find a handful of good products that you believe in and focus solely on promoting those products. You should also measure a product's demand before you start promoting it.

Use Multiple Traffic Sources

Don't rely solely on your site to drive traffic to your affiliate products. Remember, the more traffic you send the higher your income potential. Apart from your site's audience, use other sources of traffic, such as Google AdWords and social media.

Test, Measure and Track

Keep trying various promotion strategies to get traffic to your affiliate products. Do split tests, measure their performance and adjust accordingly.

Stay Up-To-Date

Affiliate marketing is highly competitive. Always stay up-to-date with new ways and techniques of promoting your affiliate products.

Choose The Right Seller

Your success as an affiliate marketer depends on your reputation. Therefore, you should only promote products from high quality sellers. If you refer visitors to a product and they end up unsatisfied with the product, they will never take your advice again, which will hurt your chances of making money in the long run.

Affiliate Program

In order to start making money as an affiliate marketer, you need to join an affiliate program. Below are two popular affiliate programs that you can join to quicken your journey to making money as an affiliate marketer.

Clickbank

This is a popular affiliate marketing program that is solely focused on digital products. It has been around for a while, being one of the first affiliate networks on the internet. Registering for an account with Clickbank is free. Once you register for an account, you get access to millions of digital products, which you can promote for a commission. One of the greatest things about Clickbank is that it pays good commissions compared to most other affiliate programs. Clickbank also allows you to advertise your affiliate products anywhere. Clickbank is very easy to use. Once you sign up for the free account, simply choose the product you want to promote, click on the promote button and receive your unique affiliate link, which you will then use to promote the product. The only disadvantage with Clickbank is that you must meet certain customer distribution and payment thresholds before you can start receiving your commissions.

Amazon Affiliates

This is the largest affiliate program on the internet. It has over 1.5 million sellers, which makes it the perfect place for you to start your journey into affiliate marketing. It is also a good choice for advanced affiliate marketers since it allows you to create custom tools and APIs for increased productivity. The high number of sellers means there is a huge selection of products to promote. The best part of becoming an Amazon affiliate is that you earn commissions on any product that the customer buys from the site. For instance, if a customer reads clicks through your affiliate link while looking for a $10 watch but also decides to buy a $1000 camera, you get commissions for both items. The Amazon Affiliate program is easy to use, flexible and very popular. The only disadvantage is that its commissions are somewhat low.

Chapter Five: Dropshipping

Dropshipping is another simple and effective way of building an online business that earns you a consistent passive income. There has been increasing interest in dropshipping in recent years and lots of people have made wealth using this online business model. Dropshipping essentially entails selling products without having to stock these products. You only need to list the products in an online store and find buyers. Once a buyer places an order for an item, you transfer the order to the supplier, who then fulfills the order and delivers the item to your customer. As a dropshipping entrepreneur, your profits are the difference between what your customer pays and what you pay to the supplier.

The beauty of dropshipping is that it helps you start a business without most of the hassles associated with starting a traditional business. You don't have to spend time and money creating your own products. You don't have to spend money leasing storage space for the products. You don't have to concern yourself with packaging and shipping either. Most importantly, you don't need tons of capital to start a dropshipping business. Instead of having to purchase hundreds of products and hoping that you will find buyers, with dropshipping you only pay for a product from the supplier after your customer has paid for the product.

As a dropshipping entrepreneur, you first need to find wholesalers or suppliers who offer dropshipping services. You then need to set up your online store and list your preferred products. You must ensure that your suppliers carry the products you intend to sell on your online store. Once your store is up and running, all you need to do is extensive marketing to get buyers to your store. When a customer makes a purchase, you will forward the confirmation email to your supplier. The supplier then takes care of the rest. The supplier packages the order and ships it to your customer. The package sent to the customer bears your business details, which means that your customer won't know that their order was fulfilled by a third party.

What Products Can You Dropship?

A dropshipping business can work with any products, in different niches and different markets. This can range from baby clothes to electronics to boats and boating equipment. However, there are criteria you can use to decide which products increase your chances of success. Ideally, you should go for products that are not expensively priced, since most people are not comfortable spending huge amounts of money online. You shouldn't choose products from generic brands, products that are confusing to use or that have multiple accessories, renewable or disposable products and products that are hard to find locally. For ease of shipping, you should also go for products that are small and lightweight. Before you decide to start selling the products on your dropshipping store, you should do adequate market research to ensure that there is enough demand for the products. You can use online platforms like Google and Amazon to measure demand for your chosen products.

Choosing A Dropshipping Supplier

There are hundreds of dropshipping suppliers. There are several ways of identifying potential dropshipping suppliers to partner with. You could get in touch with manufacturers and find out their authorized dealers. Contact these authorized dealers and find out if they offer dropshipping services. You could also search for suppliers on Google. However, most suppliers are not very intent on marketing, therefore you might need to dig deep before you find good suppliers. You can also find potential drop shippingsuppliers by attending trade shows or using supplier directories.

When searching for suppliers, you should be wary of fake suppliers who are essentially retailers selling at "wholesale prices". You also want to ensure that you have the right supplier to avoid ruining your business's reputation. Before you start working with a supplier, there are several things you should consider. How long does the supplier take to process an order and ship the product from their facility? What shipping methods does the supplier use? How do they handle issues like botched orders and lost shipments? Are there any warranties on their products? Finding out the answers to these questions is important because it helps you know how to deal with issues outside your control. Providing customer service will be much easier when you know how your suppliers handle such issues.

Dropshipping On Shopify

There are dozens of platforms you can use to run your online store. One of the simplest and most effective e-commerce platforms you can use for your dropshipping business is Shopify. Shopify is a hosted e-commerce solution that handles everything about running an online store.

Using Shopify for your dropshipping business is simple. The first step involves picking a Shopify plan that is suited to your budget and feature needs. Step two involves buying a domain name and pointing it to your Shopify e-commerce store. In step three, you get to choose a design for your online store. From there, all you need to do is add your products, fill your pages with content, set up the payment options and start looking for customers.

There are many advantages of using Shopify for your dropshipping business. Unlike hosting your own online store where you have to deal with issues like site speed and security, Shopify allows you to leave everything to the pros. Shopify will handle hackers, caching, security compliance and a whole ton of other technical issues, allowing you to focus on marketing and managing your business. Shopify also comes with all the necessary features you need for your dropshipping store, from integrated payment systems and easy inventory control to analytics and customization. Shopify also provides SEO(search engine optimization) for your site, which improves the effectiveness of your marketing efforts.

Another great Shopify feature is the ability to include the "Buy" button on multiple platforms. This means that with Shopify, people can even make purchases directly from Facebook and other social platforms, without having to visit your site. On top of that, you get great customer support, which is very crucial when you are building your very first dropshipping store.

All said and done, dropshipping is a fast, easy and low risk way of starting an online business that earns you a consistent passive income. You don't have to tie up a ton of capital in your dropshipping store. This is a business you can start with as little as $100 and turn that into a multimillion dollar empire. By building a dropshipping business, you also build a real asset. If you decide that you no longer want to be involved in the dropshipping, you can always sell your dropshipping business for top dollar.

Chapter Six: Kindle Publishing

Ever since the advent of writing, books have been a constant part of human culture. However, the advent of the internet brought about an explosion of eBooks, which have caused a massive disruption in the book publishing industry. Before the advent of the internet, publishing a book was a nightmare. You had to convince established publishers that your book was worth it in order to get published. Fast forward to today, where virtually anyone can create a name for themselves through self-publishing and create a passive income stream. Welcome to the world of Kindle publishing!

Kindle publishing involves writing an eBook and uploading it to the Amazon Kindle Store. From there, Amazon sells the eBook on your behalf, creating a nice source of passive income since you can start earning without lots of further effort from your end. The best part is that you don't have to be a writer to make money through Kindle publishing. You can make tons of money without having written a single word of the books you are selling. Before I get into how to become a publisher on Kindle, let's consider what makes Kindle publishing such an awesome opportunity.

Advantages of Kindle Publishing

Virtually Zero Barriers To Entry

One of the greatest advantages of Kindle self-publishing is that there are virtually zero entry barriers. You don't need special skills or access to any special software. You don't need to be an authority figure or even have expertise about a given topic. You don't have to do aggressive sales promotions. Heck, you can do this even without having to establish an audience beforehand.

On top of that, publishing on Kindle does not need any huge capital investments. You don't have to spend thousands of dollars to lease office space or buy any special software. Even if you are not a competent writer, you can hire a ghostwriter for your eBook for very low fees. You can spend $50 to hire a ghostwriter and then end up making $10,000 on the eBook.

Huge Market

Amazon is the king of online shopping. They are the biggest ecommerce store with over 100 million active users. By publishing on Kindle, you have access to all these potential customers. Today, the whole book industry is shifting towards eBooks. Virtually everyone in the world with an internet connection knows about Kindle. What's more, people don't even need to purchase the Kindle to be able to read your books. People can use the Kindle reader app to read your eBooks through their smartphones, tablets, laptops and desktop computers.

Another great thing about selling your eBooks on Kindle is that Amazon is a buyer's marketplace. Most of the people who visit Amazon go there with the intention of buying. This means that making a sale on the platform is incredibly easy, since most of the people on the platform are ready and willing to make a purchase.

You Don't Have To Do Outside Promotion Or Marketing

If you want, you can do your own marketing to promote your eBook and increase your sales. However, if you do not have the time or resources to dedicate to marketing your eBook, you can still leverage on Amazon's algorithms to handle your eBook's promotion. Amazon has created detailed buyer personas of its users, which means it is better placed to target the right people who might be interested in buying your book.

Selling On Amazon Kindle Store Is Passive

Another huge advantage of publishing your eBook on Kindle is that the process is generally passive. This means that you can still make money as you sleep or as you enjoy a vacation with your family. Once your book is ready and published on Amazon Kindle store, you don't need much effort to keep the money flowing in.

Writing the book, publishing it on the Amazon Kindle store and optimizing it to ensure it ranks might take some effort, but once you are done with that, you can sit back and start enjoying your profits.

With that out of the way, let's look at the actual process involved in publishing your first Kindle eBook.

Research

This should be the first step when you decide to get into kindle publishing. Doing proper market research sets your eBook up for success. Skip this step and you might just have doomed yourself to failure even before you start. One of the most common mistakes with most self-publishers is that they publish an eBook around a topic about which they are passionate or decide what to write based on intuition or some shallow keyword research on Google.

While these strategies *might* work, I do not recommend doing this. Doing that leaves your chances of success to luck. Instead, you should take the time to identify those topics that already have an established demand. Doing this greatly reduces your chances of failure because you are creating your eBook around something you are sure people want.

The good thing with Amazon is that the platform is organized in a manner that makes it possible to identify whatever is popular, whatever has the highest sales and whatever has the highest demand. Therefore, you should take the time to research on Amazon Kindle Store and identify patterns. These patterns will allow you to understand buyer behaviors, which in turn gives you insights into how the Kindle market place works and allows you to identify buyer trends before they appear.

When you are doing your research on the Amazon Kindle Store, there are a few things you should be looking out for. You should look at books focused on the niche you want to write about and find out their positions in the overall Kindle store and among their specific category. This makes it possible for you to identify the kind of books that are more likely to sell. It also allows you to find a niche that is not too crowded, one that you can easily get into and dominate.

Create An Outline For Your eBook

For purposes of non-writers who want to take advantage of Kindle publishing, I will assume that you are going to hire a ghostwriter for your book. If you are a competent writer who can write their own eBook by themselves, then do that, by all means.

Before hiring a ghostwriter to write your eBook, you need to come up with an outline that the ghostwriter will follow. The aim of the outline is to provide the ghostwriter with simple pointers on the kind of eBook you want. You don't need to go into lots of details with the outline, since this might end up stifling your ghostwriter's creativity.

The kind of outline you give your ghostwriter should be based on your market research and what the market wants. One trick you should use when coming up with the outline is to look at competitors eBooks and identify two key things – what people love and what they hate about your competitors eBooks. With that in mind, you know the kind of content readers want and the kind of content that is missing from the books available in the market.

You should also check out the table of contents in your competitors' books to get an idea of the kind of topics you need to have in your eBook and the kind of structure you should adopt for your eBook. This gets rid of any doubt you might have on what to include in your eBook. You should then create a short document with bullet points on what you want your ghostwriter to include in the eBook.

Hire A Ghostwriter For Your eBook

The beauty of Kindle publishing is that you can still make a killing even if you have zero writing skills. All you need to do is to hire a ghostwriter for your eBook. There are lots of places you can hire ghostwriters from. The most popular are sites like Freelancer, Craigslist, Fiverr and Upwork. There are other sites which are purely dedicated to eBook ghostwriting services. I will give a word of caution here. With bidding sites like Craigslist and Freelancer, there are a lot of not-so-great writers out there, so you need to make sure that you sift through the awful writers and hire only rock-star ghostwriters. There are different ways of doing this, such as checking their customer reviews and success ratings. You are more likely to get better quality with sites that are exclusively dedicated to eBook ghostwriting.

To avoid headaches after you have hired a ghostwriter, I would recommend that you give prospective ghostwriters a short test to weed out the inexperienced ones.

Create Your Title And Cover

The popular saying might claim that we should not judge a book by the cover, but no one abides with that maxim on the Amazon Kindle Store. Your book's title and cover are very important. You might have the best content in your book, but if the title and cover look whack, people won't be so enthusiastic about buying your book. On the other side, a great looking cover and a nice title will entice people to click on the book. From there, you can then use a great description to turn them into buyers.

The key to a great eBook cover and title is being unique. Let your book stand out from the rest. A unique cover catches the eye of prospective buyers as they scroll through the list of books on the Kindle bookstore. For instance, if the other book covers in your category are blue, you might decide to go with pink for your book. If you have some design skills, you can design your book's cover by yourself. If not, you can hire designers on Fiverr for just $5.

The cover is meant to make prospective buyers notice your book. From there, it is essential to convince them to check out more details about your book. It is up to your title to do this.

Here, you want a strong title that lets prospective buyers know what your book is about and makes them interested in visiting your book's individual page. If the title is unable to convince a prospect to check out your book's individual page, that is a lost sale.

Creating a strong title for your eBook boils down to copywriting. You should make use of copywriting techniques to ensure that your titles lead to a high click-through rate. Your title should speak directly to your prospects, using second-person pronouns like "you" and "your". Your title should make the benefits of reading the book very clear. What will the prospect gain by reading the book? If your book title convinces them that they will achieve whatever they want after reading your book, you are on your way to making a sale. Your titles should be blatantly simple and obvious. Don't attempt to use witty phrases or hidden meanings. The title should make it plainly clear what the book is about.

Don't be afraid to use a long title if that's what it takes to make it clear what the book is about. While short and concise titles may sound better and seem more memorable, don't use them if they cannot properly express what the book is about.

Another thing you should keep in mind when crafting your eBook's title are the keywords you want to rank for. This is a delicate affair. Don't just throw about all the keywords in your title. You should use the keywords but still maintain a natural feel in your title. Try reading the title aloud. If it sounds ridiculous, go back and change it.

Write The Kindle Store Book Description

This section also depends on copywriting techniques. Your book's description should grab your prospects' attention and hold it. While some people will buy your eBook based on the strength of your cover and title, most will need more convincing. This is the purpose of the description. Once the cover catches their eye and the title entices them to check out the individual book page, now it is up to the book description to convince them why they should buy the book.

The book description acts as the sales copy for your book. It should mention all the benefits that your prospects will gain from reading the book and help them overcome any fears they might have about purchasing the book. If possible, you should use the psychological copywriting techniques of writing a sales page when crafting your book description.

Publish

Once you get your manuscript from your ghostwriter, you should go through it thoroughly, checking its quality. If you find anything that is not satisfactory, request a revision from your ghostwriter. However, if you had done your due diligence before hiring the ghostwriter, chances are high that you will get a high-quality eBook that is ready for publishing. If the book requires only some minor changes, you can do them yourself to avoid wasting time.

Before you can publish your book, you need to format it for Kindle. You can head over to kdp.amazon.com to check out Amazon's guide and instructions on how to format your book for Kindle publishing.

Once you are done with the formatting, sign up for a free KDP account at kdp.amazon.com. This is where you will publish your book and manage everything you need to as concerns the sale of your book. Signing up for the free KDP account is easy and should take you only a few minutes.

Now is the time to publish your eBook on the Kindle store. At this point, you will be required to select a category and up to 7 keywords for your eBook. This step is very crucial since it influences your eBook's ranking on the Kindle store, which will in turn affect the number of sales you are going to make. Make sure you choose your keywords wisely. Once you are done with this step, click the preview button to view a preview of your eBook and ensure that it is properly formatted and that everything looks as it should. Some common mistakes you might find here include improper page breaking, some spacing errors and an inactive (or non-existent) table of contents.

Once you are certain that everything is okay, you can now click on "Save and Continue". This will take you to the pricing page. Here are some popular pricing models:

Free books: This means that you give away your book for free. It is a great way of building a fan base, but it won't make you any money. If it is your first time publishing an eBook on the Kindle store, you can give it away for free for the first five days to build some traction and then increase the price afterwards.

$0.99 Books: Pricing your eBook in this price range will get you lots of readers (but obviously fewer than if you give it away for free). An eBook priced in this range will make you some money, but not lots of it. This is because the royalty for books in this price range is 35%.

$1.99 Books: I tend to think that pricing your book at $1.99 is pointless. Don't do it. At this price point, the book is too expensive to attract the free and $0.99 readers. At the same time, it is too cheap to get a 70% royalty, which results in a lose-lose situation. Few readers and low profits.

$2.99 - $4.99 Books: This is the price range that allows you to make the most money from your eBook. If you want to grow your Kindle publishing business in the long term, go with a $2.99 pricing. Same case if your target audience is generally younger and poorer. If you already have an existing fan base and want to maximize your success, go with a $4.99 pricing. Same case if your prospective buyers are more affluent. If you are unable to figure out the best strategy for you, go with a $3.99 price point.

Once you decide on your book's pricing, hit publish. It will take 12-24 hours before your book goes live.

Congrats! You are now a published author.

Launch And Grow

After your book is published and live on the Kindle bookstore, now is the time for you to start making money and growing your empire.

If you correctly followed the steps outlined above, your new eBook will make some sales. However, if you want to make the big bucks, you will have to come up with methods of promoting your eBook to drive more sales.

Chapter Seven: Freelancing

The ability to work remotely through the internet and the rise of the gig economy have greatly expanded opportunities for people to make money working as freelancers. Working as a freelancer simply means working on short term contracts or assignments with different clients, companies or organizations. In a way, freelancing is like running your own business. You market your services through different channels, agree on terms of engagement with your clients and then deliver your services to the said clients.

Working as a freelancer is a great business model because it gives you the flexibility to choose your own hours and the freedom to work from anywhere in the world. All you need to work as a freelancer is a computer and an internet connection.

Ever wanted to work and keep earning as you travel throughout the world? Freelancing gives you the freedom to do just that. The great thing with freelancing is that just about anyone can do it. Everyone has some valuable and bankable skills and experiences that other people are willing to pay for.

Before I get into how you can start your freelancing business, I want to issue a disclaimer. Having a freelancing business does not mean that you will make money while you sleep. It is not uncommon to spend more time working on your freelancing business than you would have done on a traditional job. However, with freelancing, you will be in total control of your time and financial well-being and you will most likely be doing work you actually enjoy.

Advantages Of Freelancing

You are your own boss: One of the biggest benefits of working as a freelancer is that you are your own boss and therefore you don't have to put up with the stress that comes from working under a difficult person. As a freelancer, you can always drop clients who prove to be a pain in the side.

Autonomy: This is related to point number one above. Since you don't answer to anyone as a freelancer, you are free to work on your own terms, at your own rates and at your own pace. This gives you the freedom and flexibility to create your own schedules. For instance, if a relative is coming to your city tomorrow, you could finish tomorrow's work today and take the day off tomorrow.

Workload control: Another great benefit of being a freelancer is that you can control your workload. If the clients become too many for you to comfortably handle, you can simply drop one client. If you feel like you are doing too little, just pick up a new client.

Control over work relationships: As a freelancer, you can determine the number of times you want to communicate with colleagues and clients each day. For instance, you might come up with your own email policy which states that you will check and reply to emails only two times a day. Compare this to a traditional office environment where ignoring an email is akin to ignoring your boss.

Income control: Your income is not limited by your boss or anyone else. Instead, it is pegged on the amount of effort you put into your work. If you want to earn more, simply work more. There's literally no cap to your earning potential. If work becomes too much, you can outsource the extra work and keep earning.

Work-life balance: Since freelancers are location independent, they have a better work-life balance compared to people working a 9-5. You can set aside some parts of your day to spend time with your kids or to relax and do some of the things you enjoy.

Higher salary, less expenses: As a freelancer, you are in control of your earnings. Instead of asking your boss for a salary, you can increase your rates without having to consult anybody. At the same time, if you work from home, you save the money you would have spent on the commute to the office, work clothes and bought lunches each day.

Quality of work: As a freelancer, you get to choose the projects you will work on. If a client's project is not appealing or fulfilling, you can decline it. Compare this with a traditional job where you work on whatever hits your desk, whether you want to do it or not.

You learn different aspects of running a business: As a freelancer, you will essentially be running your own business. Since you are a one-man business, you will have to learn all the aspects relating to running your business, including sales and marketing, administrations, bookkeeping and so on. These are skills you might have no chance of learning in a traditional job.

How to Start Your Online Freelancing Business

To start a successful freelancing business that can comfortably sustain you, you need to do the following:

Define Your Goals

The first thing you should do is to define what you want to achieve with your freelancing business. Without a clear and measurable idea of what you want to achieve, you will have a challenging time getting there, or worse, you will end up settling for mediocre. For instance, is your freelancing business a way of earning some money on the side? Are you starting your freelancing business with the aim of becoming a full-time freelancer? Is your freelancing business a stepping stone to an entirely different goal? Whatever your ultimate goal is, make sure that it is abundantly clear. This not only applies to freelancing, but to any other business venture.

With a clear understanding of what you want to achieve from your freelancing business, it is easier to determine the minimum amount your freelancing business needs to bring in. From there, you can work out the number of clients you need and how much you need to charge them to meet your objectives.

Choose Your Craft

Nowadays, people outsource just about everything. This means that there's a very high chance that you have one or more skills that you can turn into a freelancing business. When picking a service to offer as a freelancer, you might need to think outside the box. Not everyone has technical skills. But there are several other secondary skills that people are willing to pay for. Some freelancers have built six figure businesses as dog walkers. I have even seen someone who earns over $2000 a week by standing in queues for people.

There are literally thousands of services people are willing to pay for.

Niche Down

If you want to make a lot of money as a freelancer, I would recommend that you narrow down and offer your services to a very specific niche. Let's assume you are a writer. There are millions of other people providing freelance writing services on the internet. If you want to stand out from this crowd, you have to narrow down and position yourself as an expert in a specific niche. For instance, you might decide that you will only offer your services for B2B tech companies. If a B2B tech company wants to hire a writer, who do you think they will hire, you or a general writer who writes just about anything? They will hire you of course, because you have positioned yourself as someone who is knowledgeable in that particular industry. On top of that, niching down allows you to command higher rates because you have already established yourself as an expert in that particular niche.

Build A Portfolio And Look For Testimonials

One of the things that differentiates the world of freelancing from the corporate world is the lack of adherence to useless rules and formalities. In the freelancing world, clients do not care about your qualifications. They only want to know that you can deliver results. The only way they can be sure that you *are* capable of delivering results is by seeing what you have done in the past. That is why it is very important for every freelancer to have a portfolio. A portfolio helps demonstrate your skills to potential clients. You should also collect testimonials from previous clients showing what these clients thought of your services. A solid portfolio and positive client testimonials greatly increase your chances of success as a freelancer. To build up your portfolio, don't be afraid to do *pro bono* work for your ideal clients. This helps showcase your skills in better light as compared to doing cheap work which then sets you in a vicious cycle of low paying work. Before a client agrees to pay you big money for your services, they want to be sure that you can deliver work that is worth the big money.

Price Your Services Strategically

There are millions of people offering freelance services on the internet. This means that there will always be people willing to provide the same services you do at a lower price. If you decide to compete with these people based on price, you will soon be working yourself to death at rates that can barely pay your bills. Instead of charging what your competitors are charging, price your services based on the value you provide. Many freelancers make the mistake of being afraid to charge prices that are too high. However, there is nothing like a price that is too high. If you provide a valuable service and target the right clients, you will get clients even if you charge high rates.

Start Pitching

Once you have defined your goals, decided on your craft and niche, created a portfolio and determined the rates you are going to charge for your services, now is the time to start pitching to clients. There are several places where you can find potential clients. The most popular places for finding potential clients are freelancing websites, which I am going to discuss in further detail later. When pitching, you should keep two things in mind – relevance and volume. This means that you should send *lots* of pitches but *only* to clients within your specific niche. With such an approach, you will easily find clients for your freelancing business.

10 Highest Paying Skills For Freelancers

I mentioned earlier that there are hundreds of skills which people are willing to pay for. However, there are some freelancing skills which are in high demand and which in turn command high rates. These include:

Programming And Software Development

This is the highest paying skill for freelancers. Becoming a good programmer or software developer is quite a challenging skill to master. There are not many skilled programmers and developers out there, therefore the demand for this skill is quite high. If you have solid coding skills, you can easily make a lot of money as a freelancer. This does not mean that you can start your freelancing business today and start charging $1000 per hour tomorrow. You will need to grow your reputation gradually. However, it is possible to start to earn about $150 per hour even if you are a beginner.

Web Design And Development

According to a statistics report by Upwork, PHP development is one of the skills with the highest demand on the platform. These days, virtually every business has a website, which explains why there is such a huge demand for web designers and developers. However, web design and development is not a very difficult skill to master, which means that the web design and development industry is a bit crowded. However, you can still make good money as a freelance web designer, especially if you know how to differentiate yourself and go the extra mile to provide more value for your clients.

Content Writing And Marketing

Lately, the online marketing industry is experiencing a boom in new trends like inbound marketing. As more and more businesses try to lure prospects to their websites with inbound marketing, there has been an explosion in the demand for content writers. Anyone can make money as a writer. You don't need a literature degree either. However, you need to be a highly skilled writer to make good money. You will need to read and write a lot and generally be very creative. If you are highly skilled in your craft, you can easily make over $5000 a month as a freelance writer.

Graphic Design

This is another freelancing skill that is in high demand. Think of all the visual marketing material that is put out by businesses every day. All this material is created by graphic designers; therefore, you can see why this skill is in such high demand. Think of brochures, logos, posters, infographics, icons and illustrations. As a freelance graphic designer, you can easily make around $85 per hour. However, you need to have a keen eye for design and become pretty good in your craft to stand out, since this industry is quite crowded.

Copywriting

Though this is also a writing skill, it is different from content writing. Copywriting involves writing content whose aim is to sell products and services. Copywriters write the copy for sales pages, product descriptions, lead magnets and squeeze pages and so on. The amount of money you can make as a copywriter depends on your skill and experience. Highly experienced copywriters earn up to $250 per hour. However, the typical rate for most copywriters ranges between $15 and $100 per hour.

Video Editing

There has been increasing demand for video content due to the opportunities presented by social platforms such as YouTube, Vimeo, DailyMotion, Facebook and Instagram. This has led to a spike in the demand for video editors. If you can put together video clips and come up with awesome video content, that is a great skill on which you can base your freelancing business. Freelance video editors in the US earn anywhere between $72,000 and $122,000 per year according to research by SimplyHired.

Social Media Management

Nowadays, social media platforms have become more than social tools. They have become marketing tools for businesses to promote themselves. Today, almost every business, from local eateries to big corporate brands use social media networks as part of their brand and business marketing strategies. Most of these businesses do not have the necessary skills to run effective social media marketing campaigns, which creates an opportunity for freelancers to make some money managing social media pages for businesses. If you know the ins and outs of social media, you can easily make money managing a business's social media pages.

Voice Acting

If you have a smooth and wonderful voice, you can earn some money on the side recording voiceovers. Since many businesses are now using video as part of their marketing strategies, this is a skill that will continue to see rising demand. You can make up to $72 per hour as a freelance voiceover artist.

Search Engine Optimization

Search engine optimization (SEO) involves optimizing websites to ensure that they can be easily found on search engines. Many businesses want their sites to rank on search engines but do not have the necessary skills to do this. You can learn this skill in a short time and start earning up to $50 per hour.

Translation

If you happen two know two or more languages, you can make money translating between the two languages. Spanish-English and Korean-English translations pay particularly well. However, you can still make money translating between many different languages. Depending on your skills, nature of the task and the languages involved, you can make between $25 and $40 per hour as a translator.

The above are the top ten highly paying freelancing skills. However, this doesn't mean that you can't make it as a freelancer if you don't have one of the above skills. There are several other skills that you can use to start your freelancing business.

The Best Freelancing Websites

Freelancing websites are online platforms that aim to connect freelancers with clients. As a freelancer, you can save yourself a lot of time and effort by finding your clients on freelancing sites, since you are sure the clients on these sites are looking for people to help them with a specific service. Some popular freelancing sites include:

Upwork: This is the world's largest and most popular freelancing websites. It was formed after a merger between two leading freelance networks, Elance and Odesk. Upwork has over four million registered clients and over 3 million job postings annually. There are all kinds of freelancing jobs posted on Upwork.

Freelancer: This is another big and popular freelancing marketplace, second only to Upwork. Freelancer also has millions of small businesses that are looking for freelancers. However, you should be careful when looking for jobs on Freelancers, because it is possible to find jobs that are posted by middlemen instead of the actual clients.

Fiverr: This freelancing marketplace is a bit different from the others. Here, freelancers post services (known as gigs) which thy can offer for $5, hence the name Fiverr. However, this does not mean that you must sell your services for $5. You can offer additional perks to your base service and charge much higher prices.

PeoplePerHour: This is a site dedicated to freelancers who provide services related to web projects, such as software engineering, SEO, online marketing and UX design. The great feature about PeoplePerHour is that it has a centralized feature that handles the business side of your freelancing.

SimplyHired: This is another large freelancing network that provides all kinds of freelancing opportunities, from concierge work to construction.

If you have some time to spare or if you are enticed by the prospect of setting your own hours and working from anywhere in the world, you should consider starting a freelancing business.

Chapter Eight: Selling Online Courses

There's a wide variety of reasons why people surf the internet. A wide majority of people go online for entertainment and to socialize with others. However, an increasing number of people are going online search for information and to learn new things. If you want to learn how to use Excel or Photoshop or WordPress, there's probably a video tutorial on YouTube explaining how to do just that.

This search for knowledge and learning on the internet has created an opportunity for people to make money from the knowledge they have by creating blogs and writing eBooks. A new way of making money from the knowledge you have is through sale of online courses. An online course is basically information packaged in a manner that helps someone to learn how to do something through a series of lessons.

The good thing about making money from selling online courses is that anyone can create an online course and make money from it. Everyone has enough knowledge about something that others will find worth paying for. You don't even need to be an expert. You only need to know more about a topic than the average guy to be able to create an online course. Online courses are also not confined to any topic. You can make money teaching people about anything, from speaking a new language to cooking to playing an instrument. I have even seen guys make money from teaching others how to kiss.

Why You Should Start Selling Online Courses

Before we go into how you can make money selling online courses, let's look at why you should start creating and selling online courses.

Online courses provide you with a great means of earning a passive income. Once you create your online course and publish it online, you can sit back and start making money in your sleep. Of course, if you want to make good money you will need to promote your online course.

Online courses are digital products, which means they are highly scalable. Most online courses are in video format. Students pay to access and view these videos. Once you have created your course and sold it to one student, it doesn't cost you anything to sell it to another student. Even if 10,000 students purchase the course at once, there's no extra effort on your part. All you need to do is collect your money. Compare this with a live class. If your students increased from 20 to 100, you would need to move to a bigger classroom and perhaps divide the students into two or three groups.

According to Forbes, the online learning market is expected to exceed $240 billion by 2023. As more and more people get access to the internet, more people are willing to purchase online courses. Udemy, one of the leading online course marketplaces has 9 million registered students. Since this market is on an upward trend, it means the time is still right to tap into this market. Additionally, more and more people are choosing the convenience of online courses over traditional classrooms.

Today, there are several online course marketplaces where you can sell your online course. These platforms make the whole process easy by providing you with all the necessary tools you require to sell your online course. On top of that, they also give you access to their traffic. For instance, if you sell your online course on Udemy, the marketplace gives you access to its 10 million+ students. Some of these marketplaces will even drive students to your course, which means you can still make money even if you are poor at marketing and promotion. Additionally, since your course is available online, there are no geographical barriers to the number of people you can reach. You can sell to people from any part of the globe.

Most people are held back from selling online courses by the fact that the information they intend to sell through their course can be found for free online. While this is true, people are willing to pay for a course that has distilled this information together in a clear manner instead of spending hours scouring the entire internet for tutorials. On top of that, people are willing to pay for online courses because they get exclusive access to an instructor who can answer their questions.

Online courses can be offered as part of an existing business. For instance, if you are a digital marketing consultant, you can offer a course that teaches people how to do digital marketing for their businesses.

How To Create Your Online Course

Creating your own online course can be divided into the following 10 steps:

Step 1: Choose A Topic For Your Course

As I mentioned earlier, to create an online course, you need to know more about the subject than the average person. To come up with a topic for your course, make a list of things you are good at. If you have a hard time figuring this out, try identifying some of the things with which your family and friends come to you for help. You could also ask them to help you identify things you are good at. Your hobbies are another great source of topic ideas for your online course.

Step 2: Market Research

Before you set out to create an online course, you want to know if it is something people want to learn or you will waste your time creating a course that no one will buy. In some cases, people might be curious about a topic but still be unwilling to learn about it. Therefore, when doing your market research, identify the kind of people your course is targeting and find out whether they are ready, willing and able to buy your course.

Step 3: Create An Outline For Your Course

Once you are sure that there are people who are ready, willing and able to pay for your course, now is the time to figure out what you to include in your course. When creating a course about a particular topic, you should provide highly detailed content that covers all the important elements in that topic. Do not create a course that provides shallow content that would fit in a blog.

For better organization, arrange the content in terms of modules and lessons. In this case, a module should combine several lessons that talk about related subtopics.

Step 4: Decide On How You Are Going To Deliver Your Lessons

Great online courses come in a variety of teaching methods. For instance, you might decide to provide your lessons in video, audio or text formats, with worksheets, infographics, checklists and anything else you feel is necessary to help your students understand the lesson better. There is no perfect delivery method for an online course. It all depends on what you are trying to teach. For instance, if you are teaching people how to use Excel, you can either use video or text combined with images. However, teaching someone how to play the guitar through text might be a bit challenging. If possible, you should provide multiple delivery methods for each lesson.

Step 5: Create The Actual Lessons

Creating the actual lessons for your online course is the bit that will take the most time and require the most effort. For branding purposes, consider having a consistent theme for all your lessons. Once you are done creating the lessons, proofread them and watch the lesson videos to ensure that everything is as you want it.

Step 6: Determine How You Will Sell Your Online Course

This is the part where you decide on how you are going to put your course in front of prospective students. If you want to have utmost control over your course and the entire learning experience, you should create your own website and host the course by yourself. There are dozens of WordPress plugins and membership site scripts that you can use to simplify the process of selling and delivering your online course. However, hosting your course on your own website might require you to have some technical skills.

If you don't have the time and technical skills required to host your online course on your own website, you might opt to use online course platforms. There are several online course platforms on the internet. Some of the most popular include:

Udemy: This online course marketplace is one of the most popular for students who want to learn some new skill. With over 10 million students, you are assured that you will get students interested in your course. Udemy allows instructors to create courses about any topic. The platform also provides instructors with advanced but user-friendly design tools which you can use to create your online course. Many people selling online courses prefer Udemy because of its user-friendliness, flexibility, high earning potential and the round-the-clock customer service. Popular courses offered on the platform include photography and design, programming, software development and online marketing.

Teachable: This is another popular platform that makes it easy for educators to create and sell their online courses. Teachable has an easy and intuitive central interface that helps you create a well-designed and responsive site website. Through this central interface, you can also customize your website to suit your brand, manage student data and determine the pricing for your online course.

You don't need any coding skills to be able to use Teachable. The platform handles the hosting, student sign-ups, payments and back-end aspects of your site, allowing you to seamlessly customize and launch your course website. Signing up to sell your course on Teachable is free. However, you can upgrade to a paid plan to increase the platform's functionality.

Skillshare: This platform is based on the premise that everyone has something they can share with the world. It provides people with the right tools to create classes and reach out to students from all over the world. Just like Udemy, Skillshare does not have restrictions on the kind of courses that educators can offer. Some of the commons courses on Skillshare are on subjects like design and creativity, fashion, writing, technology, gaming, photography and film, fashion, arts and crafts, culinary arts and so many more.

Ruzuku: One of the best things about this platform is that they are very dedicated to helping educators use the platform effectively. This includes everything from creating a great course and getting it online to promoting and marketing it. Another great thing is that it has a paid plan that provides educators with unlimited webinars. Ruzuku is a great option if you are looking for an easy-to-use but highly functional platform. You don't need any technical skills to host your online course on Ruzuku. On top of that, it also has some other great features like daily backups and PayPal and MailChimp integrations.

CourseCraft: This is a platform that is geared towards helping people create e-courses and turn their blogs into actual income generating businesses. Like the other online course platforms mentioned above, CourseCraft is easy to use. It even allows you to view other live online courses that have been made using the platform.

Most of the platforms highlighted above don't require any exclusivity, which gives you the freedom to sell your online course on multiple platforms. However, you should read their terms of service before you start selling your course on multiple platforms.

Step 7: Publish Your Course Online

After you have created your online course and decided on the most appropriate platform to host it on, you can now go ahead and upload your course on the platform. At this point, your course is live and students can now start enrolling.

Step 8: Marketing Your Online Course

If you want to make a serious income from your online course, you will need to promote and market it, regardless of the platform you are using. Come up with a marketing plan to promote your online course. Your marketing plan should identify your target market, how to reach them and how to convince them to try your course. Some methods you can use to promote your course include social media marketing, article marketing and PPC ads.

Step 9: Keep Updating Your Course

If you want to keep earning from your online course for a long time, you should keep updating it every few months to ensure that the information in the course is current and relevant. Outdated information might lead to bad review, which will drive down your sales and damage your reputation.

Step 10: Rinse And Repeat

You can publish as many online courses as you want. Ideally, you should create online courses on related topics. This way, you can refer students from one course to your other courses.

Online courses are a great way to earn a passive income, provided you are able to provide great content and effectively promote your courses. While creating the course content might be tedious and time consuming, once you publish your course, you don't need much effort to keep the money flowing in

Conclusion

Thanks again for taking the time to purchase this book!

You should now have a good understanding of passive incomes and be able to earn a passive income by starting your own online business.

If you enjoyed this book, please take the time to leave me a review on Amazon. I appreciate your honest feedback, and it really helps me to continue producing high quality books

About The Author

31-year-old Anthony Parker is a self-made Internet entrepreneur, investor & author.

Anthony first entered the world of Online Business in 2014 with the creation of his first dropshipping store, since then Anthony has created and sold 3 other dropshipping stores and has since ventured out into Amazon FBA and Affiliate Marketing with the goal of creating long-term highly profitable passive income streams.

Anthony latest venture is to share his knowledge and passion on the world of Online Business with the goal of making seemingly complex and intimidating topics simple and easy-to-read with the hope of encouraging others to become internet entrepreneurs.